"Well, the passage of time has revealed to everyone the truths that I previously set forth; and, together with the truth of the facts, there has come to light the great difference in attitude between those who simply and dispassionately refused to admit the discoveries to be true, and those who combined with their incredulity some reckless passion of their own.

Men who were well grounded in astronomical and physical science were persuaded as soon as they received my first message. There were others who denied them or remained in doubt only because of their novel and unexpected character, and because they had not yet had the opportunity to see for themselves. These men have by degrees come to be satisfied."

        **Galileo Galilei: Letter to the
        Grand Duchess Christina
        of Tuscany, 1615**

All rights reserved under article two of the Berne Copyright Convention (1971).
We acknowledge the financial support of the Government of Canada through the
Book Publishing Industry Development Program for our publishing activities.

Published by Apogee Books, Box 62034, Burlington,
Ontario, Canada, L7R 4K2, http://www.apogeebooks.com
Tel: (905) 637-5737

Printed and bound in Canada

Deep Space Pocket Space Guide by Stephen Whitfield
Apogee Books Pocket Space Guide #3

ISBN-10: 1-894959-29-9
ISBN-13: 978-1-894959-29-2

©2005 Apogee Books

DEEP SPACE

Pocket Space Guide

by Stephen Whitfield

Deep Space - Pocket Guide

Exploring Deep Space

For as long as mankind has possessed a sense of wonder, astronomy has been practiced in one form or another. It can be thought of as having graduated to the status of a true science at the point in history when it became forever separated from astrology. Clearly they were not of the same mold; astronomy was hard to do; astrology was merely hard to swallow.

The next significant change came in the early 20th century. Men and women of science felt it was no longer sufficient to simply catalog the visual characteristics of the stars and other heavenly bodies; the desire to understand their physical makeup, behavior and creation gave us the study of astrophysics. As the name suggests, this is the physics of astronomical bodies and the space they inhabit.

During the second half of the 20th century interest turned also to the other eight planets of our solar system, spawning several new disciplines relating to Planetology, the formation of the solar system, and the possibility of past or present extraterrestrial life. Every decade of the last century brought ever more numerous and fascinating discoveries. It was not until 1930 that Pluto, the ninth planet, was discovered. Only as late as 1978 was it discovered that Pluto possessed a moon. And now, in late 2005, NASA's Hubble Space Telescope has given astronomers evidence that Pluto may have not one, but three moons.

In recent decades much interest has also been given to the far reaches of the universe – far stars; other galaxies, some old, some still forming; and even the earliest detectable galaxies of the universe, at the very limits of our ability to probe. Each field of space study requires its own special tools and its own expertise. The many programs that come loosely under the headings of astronomy and space exploration are as varied and rewarding as the fascinating discoveries that they have made. To the uninformed, the exploration of deep space may seem esoteric and unexciting, but in truth the wonder and the beauty of deep space outshine almost anything that Earth and man can offer.

The time when people can go and explore deep space in person is still well into the future. Until then, we shall continue to do as we have done for the past 40 years, sending unmanned spacecraft and satellites as our proxies and ambassadors to the universe. This book examines deep space exploration missions, past, present and future, and their results.

How Deep is Deep Space?

The term "deep space" tends to be a relative term rather than an absolute designation. The greatest distance that men have traveled is to the Moon, which is 384,402 km (238,856 miles) from the Earth — right next door by astronomical standards. The farthest that men have currently seriously contemplated traveling is to Mars, a little over 200 times farther from Earth than the Moon is. Trip times to Mars are typically quoted at six to eight months, one way.

In July of 1965, the *Mariner 4* Mars flyby spacecraft returned the first-ever close-up photos of another planet. Since then, more than a dozen spacecraft have traveled to Mars, and future (unmanned) Mars missions are planned as far ahead as the year 2013. There are still many advocates of a manned Mars mission within the next two decades. By comparison, only six missions have been sent to planets farther from Earth than Mars — *Pioneer 10*, *Pioneer 11*, *Voyager 1*, *Voyager 2*, *Galileo* and *Cassini*.

All of this taken together would suggests that we can reasonably use the orbit of the planet Mars as a boundary line. In this book we shall define "deep space" as anything further from the Sun than the orbit of Mars.

If we use the flight paths of *Voyager 2* (Earth-Jupiter-Saturn-Uranus-Neptune) and *Cassini* (Earth-Jupiter-Saturn) to calculate spacecraft speeds within the solar system with current technology, we get an average speed of just over half a year per AU (astronomical unit). More conveniently expressed, spacecraft speeds in the solar system average about 2 AU/year. This value is, of course, dependent on the relative positions of the planets at launch, gravity assists, propulsion technology, and more, but it will serve as a reasonable yardstick for comparing planetary journeys. If we look at orbit-to-orbit (not necessarily planet-to-planet) trips within the solar system using 2 AU/year, we get the following table:

	AU from Earth Orbit	**Travel Time in Years** *
Jupiter	4.20	2.1
Saturn	8.54	4.27
Uranus	18.22	9.11
Neptune	29.06	14.53
Pluto **	38.50	19.25

 * These values average about 10% longer than *Voyager 2* achieved, but it took advantage of a rare planetary alignment which gave it an optimum trajectory to the outer planets.

** This is an average value. Pluto's orbit is extremely elliptical.

So, from Earth to Pluto and back, the travel time alone would be half a human lifetime. It brings home the words of John W. Campbell, Jr., who said that space exploration would have to await something better than rockets. Clearly, for the foreseeable future, deep space exploration will be the province of unmanned spacecraft. A professional person attached to a deep space mission may well face dedicating most or even all of an entire career to a single mission. To study the universe outside our solar system we must rely on astronomical telescopes in their various configurations.

Part 1 – The Inner Solar System

Some past space missions considered as deep space missions fall short of our Mars orbit boundary (1.52 AU from the Sun) but are worth mentioning because of the distances from Earth at which they accomplished significant events. This section discusses these missions in order of increasing distance from the Sun.

GIOTTO – *Giotto* was Europe's first deep space mission. Among its achievements were: first close-up images of a comet nucleus, including Comet Halley; first spacecraft to encounter two comets; and first deep space mission to change orbit by returning to Earth for a gravity assist.

Giotto passed within 596 kilometers of the nucleus of comet Halley while it was 0.89 AU from the Sun (0.98 AU from Earth) in March 1986. *Giotto* flew to within a mere 100 to 200 kilometers of comet Grigg-Skjellerup while it was 1.01 AU from Sun (1.43 AU from Earth) in July 1992, after which the mission was designated as completed.

Giotto was also noteworthy for having survived the dust impacts of the Halley encounter (which was unexpected), and for having been put into "hibernation" for an extended period and then "reawakened" for navigational corrections and its second cometary encounter.

NEAR-SHOEMAKER – The Near Earth Asteroid Rendezvous (NEAR) mission, also called NEAR-Shoemaker, was the first of NASA's Discovery missions, and the first mission ever to go into orbit around an asteroid. It was launched in February 1996 on a Delta II launch vehicle.

In June 1997 *NEAR* flew within 1,200 km of the main belt asteroid 253 Mathilde. During the flyby *NEAR* took pictures with a resolution of 180 meters/pixel. The images were used to study the size, shape, surface features and colors of

Mathilde, and to search for any small moons it may have had. *NEAR* was 1.99 AU from the Sun (2.19 AU from Earth) when it flew past asteroid 253 Mathilde.

NEAR's primary mission was to rendezvous with and achieve orbit around the near-Earth asteroid 433 Eros in January 1999, and study the asteroid for approximately one year. Technical problems resulted in a delay to the mission plan. The revised mission saw *NEAR* perform a flyby of Eros in December 1998, and then finally rendezvous and orbit Eros in February 2000. At this time, *NEAR* and asteroid 433 Eros were 1.16 AU from the Sun and 2.63 AU from Earth.

NEAR touched down on the surface of Eros on February 12, 2001 and contact was maintained. Sufficient electrical power was available to transmit data at 10 bits/sec. The spacecraft obtained 69 high-resolution images before touchdown, and impacted at a speed between 3.4 and 4.0 mph. The last image obtained showed an area about 6 meters (20 feet) across. The last communication with the *NEAR* spacecraft was on February 28, 2001, as it sat on the surface of Eros.

DEEP SPACE 1 – NASA's New Millennium program, managed and operated out of JPL, was initiated in the mid-1990s for the purpose of testing new space-related technologies. The Deep Space 1 project was the first of the New Millennium missions, a spacecraft to test advanced technologies, including:

Ion propulsion system
solar concentrator arrays
autonomous navigation and flight software
beacon monitor
miniature camera / spectrometer
plasma instrument
small deep-space transponder
Ka-band solid-state power amplifier
low-power electronics

Launched in October 1998, *Deep Space 1* was propelled by a xenon-ion engine and hydrazine-fueled thrusters. The ion engine was used to add thrust as the spacecraft orbited the Sun.

In July 1999 *Deep Space 1* flew within an estimated 16 miles (26 km) of asteroid 9969 Braille (formerly 1992 KD) at a distance of 1.34 AU from the Sun and 1.26 AU from Earth. Following the end of its primary mission in September 1999, *Deep Space 1* went on an extended mission, flying by comet Borrelly in September 2001, obtaining the best-ever images of a comet's nucleus. The spacecraft was retired in December 2001.

DEEP IMPACT – Launched in January 2005, NASA's *Deep Impact* spacecraft reached the vicinity of comet Tempel 1 in July 2005, for the purpose of making measurements to compare the makeup of the comet's surface with that of its coma and interior.

At a distance of 1.50 AU from the Sun and 0.89 AU from Earth, the *Deep Impact* spacecraft released a smaller "Impactor" spacecraft into the path of comet Tempel 1. The main "flyby" spacecraft then observed the collision of the impactor with the comet. When the impactor hit the sunlit side of comet Tempel 1, there was a brilliant and rapid release of dust that momentarily saturated the cameras onboard the spacecraft. All available orbiting telescopes watched from space, including the Spitzer, Hubble and Chandra telescopes.

The impactor's impact with comet Tempel 1 formed a crater, with ice and dust debris ejecting from the crater revealing freshly exposed material offering clues to the early formation of the solar system. The comet was found to have a very fluffy structure that is 'weaker than a bank of powder snow.' The fine dust of the comet is held together by gravity. The comet nucleus appears to be extremely porous, which allows the surface of the nucleus to heat up and cool down almost instantly in response to sunlight. If this keeps heat from being easily conducted to the interior, the material deep inside the nucleus may be unchanged from the early days of the solar system.

Sixteen days after it's encounter with comet Tempel 1, the Deep Impact team placed the spacecraft on a trajectory to fly past Earth in late December 2007. The maneuver allows NASA to preserve options for future use of the spacecraft.

The amount and brightness of the released debris indicates that beneath the surface of the comet there is microscopic dust; water and carbon dioxide ice; and hydrocarbons. Signatures of these species were seen in spectra immediately after impact.

STARDUST – Stardust is another of NASA's Discovery missions, this one launched in February 1999, to fly through the cloud of dust that surrounds the nucleus of comet Wild-2 (pronounced 'Vilt-2') and return the cometary material to Earth for study. This will be the first-ever extraterrestrial material returned to Earth from outside the orbit of the Moon. The sample is due to be returned to Earth in January 2006.

Wild-2 is a 'well-preserved" comet, which makes it of special interest to astrobiologists because it represents samples of the 'fundamental building blocks of the solar system." Ablating inner-system comets provide Earth with a source of organic material and water from the outer regions of the solar system. Most of the atoms in our planet, and our in bodies, derive from particles released by comets like Wild-2.

The coma (head) and tail of a comet are composed of gas and dust that are liberated from the comet by solar heating. Rather than trailing the comet, as would happen on Earth, the comet's tail always points away from the Sun, because the solar wind is the only force acting on the liberated gas and dust.

The science-related subsystems carried by Stardust include:

Aerogel dust collectors
sample return capsule
comet and interstellar dust analyzer
dust flux monitor
navigation camera

In January 2004, after five years in space, Stardust experienced its brief encounter with Wild-2 at a distance of 1.85 AU from the Sun and 2.6 AU from Earth. The spacecraft flew within 147 miles (236 km) of the comet, impacting millions of dust particles and small rocks (up to 5 mm diameter) at high speed. Stardust captured thousands of comet particles for return to Earth. Over the entire mission, Stardust will have flown a total of about 3.2 billion miles (5.2 billion km).

Stardust also took pictures during its flyby. Among other startling sights, the images showed sunlight reflecting off more than jets of dust projecting into space, produced when gas escapes from localized regions and carries dust and rocks outwards. Dust detectors on the spacecraft measured bursts of impacts when Stardust flew through these jets.

DAWN – NASA currently has plans for a 2006 launch of its *Dawn* spacecraft. Dawn's goal is to better understand the conditions and processes of the very early solar system. It will investigate the internal structure, density and chemical makeup of two protoplanets – Ceres and Vesta – which are very different in nature, but both have remained intact since their formation. Dawn will be the first spacecraft ever planned to orbit two different bodies, in two very different orbits, after leaving Earth.

Dawn's Major Measurements Objectives:

> Internal structure, density and homogeneity of two complementary protoplanets (1 Ceres and 4 Vesta), one wet and one dry
> Determine shape, size, composition and mass
> Surface morphology, cratering
> Determine thermal history and size of core

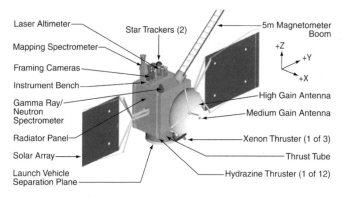

The *Dawn* spacecraft.

Ceres and Vesta both reside in the asteroid belt between Mars and Jupiter. However, they have followed very different evolutionary paths and are correspondingly different in their current conditions. Ceres may have active hydrological processes, leading to seasonal polar caps of water frost; it may also have a thin, permanent atmosphere, distinguishing it from the other minor planets. Vesta may have rocks that are more strongly magnetized than those of Mars. *Dawn's* discoveries could tell us much about the nature of the building blocks from which the terrestrial planets were formed.

ROSETTA –Launched in March 2004, ESA's *Rosetta* spacecraft will perform the first-ever long-term, close-up exploration of a comet. The spacecraft will enter into orbit around Comet 67P/Churyumov-Gerasimenko in 2014, at which time its included small lander will descend to the comet's nucleus. When it achieves cometary orbit *Rosetta* will be 5.35 AU from the Sun and 6.68 AU from Earth.

The *Rosetta* orbiter has eleven scientific instruments:

- Ultraviolet Imaging Spectrometer
- Comet Nucleus Sounding
- Cometary Secondary Ion Mass Analyser
- Grain Impact Analyser and Dust Accumulator
- Micro-Imaging Analysis System
- Microwave Instrument for the Rosetta Orbiter
- *Rosetta* Orbiter Imaging System
- *Rosetta* Orbiter Spectrometer for Ion and Neutral Analysis
- *Rosetta* Plasma Consortium
- Radio Science Investigation
- Visible and Infrared Mapping Spectrometer

The *Rosetta* lander has nine scientific instruments:-

- Alpha Proton X-ray Spectrometer
- *Rosetta* Lander Imaging System
- Comet Nucleus Sounding
- Cometary Sampling and Composition experiment
- Evolved Gas Analyser
- Multi-Purpose Sensor for Surface and Subsurface Science
- RoLand Magnetometer and Plasma Monitor
- SD2 Sample and Distribution Device

Surface Electrical and Acoustic Monitoring Experiment, Dust Impact Monitor

The Microwave Instrument on the *Rosetta* orbiter was supplied by the Jet Propulsion Laboratory. It will study gases given off by the comet while the spacecraft orbits it.

Rosetta's journey to Comet 67P/Churyumov-Gerasimenko will last 10 years. The mission will ends December 2015 when the spacecraft will again pass close to Earth's orbit.

SUMMARY OF INNER SOLAR SYSTEM MISSIONS

Spacecraft	Destination	From Sun
Giotto	Comet Halley	0.89 AU
Giotto	Comet Grigg-Skjellerup	1.01 AU
NEAR-Shoemaker	Asteroid 433 Eros	1.16 AU
Deep Space 1	Asteroid 9969 Braille	1.33 AU
Deep Space 1	Comet Borrelly	1.34 AU
Deep Impact	Comet Tempel 1	1.50 AU
	Mars orbit	1.52 AU
Stardust	Comet Wild-2	1.85 AU
NEAR-Shoemaker	Asteroid 253 Mathilde	1.99 AU
Dawn	Vesta (protoplanet)	2.36 AU
Dawn	Ceres (protoplanet)	2.77 AU
	Jupiter Orbit	5.20 AU
Rosetta	Comet 67P/Churyumov-Gerasimenko	5.35 AU

Part 2 – The Planets

JUPITER

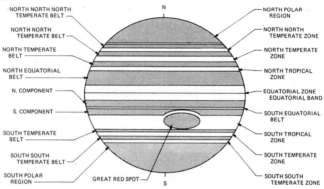

The visible surface of Jupiter consists of a banded structure of clouds that are consistent enough to have been given the names shown in this drawing.

JUPITER – Facts & Figures
The orbit of Jupiter is 4.2 AU from Earth.
Jupiter has 63 known moons.
Average Distance from the Sun: 778,412,020 km / 483,682,810 miles / 5.20336 AU
Perihelion (closest approach to the Sun):   740,742,600 km / 460,276,100 miles / 4.952 AU
Aphelion (farthest distance from the Sun):   816,081,400 km / 507,089,500 miles / 5.455 AU
Equatorial Radius: 71,492 km / 44,423 miles (11.209 × radius of Earth)
Equatorial Circumference: 449,197 km / 279,118 miles
Volume: 1.4255×10^{15} km^3 / 3.42×10^{14} miles^3 (1,316 × Earth's volume)
Mass: 1.899×10^{27} kg (317.82 × Earth's mass)

JUPITER – Facts & Figures
Density: 1.33 grams/cm^3 (0.241 × Earth's density)
Surface Area: 6.218 × 10^{10} km^2 / 2.401 × 10^{10} miles^2 (121.9 × Earth's surface)
Equatorial Surface Gravity: 20.87 m/sec^2 / 68.48 ft/sec^2 (2.14 × Earth weight)
Escape Velocity: 214,300 km/h / 133,200 mph (5.33 × Earth's escape velocity)
Sidereal Rotation Period (Length of Day):   0.41354 Earth days / 9.925 hours
Sidereal Orbit Period (Length of Year):   11.8565 Earth years / 4330.6 Earth days
Mean Orbital Velocity: 47,051 km/h / 29,236 mph (0.0439 × Earth orbital velocity)
Orbital Eccentricity: 0.04839 (2.90 × Earth's orbital eccentricity)
Orbital Inclination to Ecliptic: 1.305 degrees
Equatorial Inclination to Orbit: 3.12 degrees (0.0178 × Earth's equatorial inclination)
Orbital Circumference: 4,774,000,000 km / 2,996,000,000 miles (5.165 × Earth)
Effective Temperature: -148 °C / -234 °F / 125° K

Pioneer 10 – Man's first spacecraft to Jupiter, reached the giant planet in December 1973, after a 620-million-mile (998-million-kilometer) journey that took nearly two years. *Pioneer 10* returned color pictures of Jupiter and its inner moons, and provided information on its turbulent atmosphere; murky interior; and magnetic, electrical, and radiation environment. A few hours before its closest approach to Jupiter, *Pioneer 10* encountered its intense radiation belts. After Jupiter, the spacecraft continued into deep space, and out of the solar system.

Prior to the arrival of *Pioneer 10*, most of what was believed about Jupiter was conjecture. With *Pioneer 10's* findings, Jupiter became a great deal better understood. Based on *Pioneer 10* data, the following was learned about Jupiter:

Jupiter is almost certainly a liquid planet, for it is too hot to solidify, even with its enormous internal pressures of millions of atmospheres.

Both temperature and pressure rise with depth.

At the transition zone to liquid, 1,000 kilometers (600 miles) below the top of the atmosphere, the temperature is calculated to be 2,000° C (3,600° F).

At 3,000 kilometers (1,800 miles) down, the temperature is 5,500° C (10,000° F) and the pressure is 90,000 Earth atmospheres.

At 25,000 kilometers (15,000 miles) down, the temperature reached 11,000° C (20,000° F) and the pressure is three million atmospheres. At this level, liquid hydrogen turns to liquid metallic hydrogen.

Unlike Earth, Jupiter has no concentrations of mass, such as a rigid crust or other solid areas. Measurements show that the planet is in "hydrostatic equilibrium," or largely liquid.

Jupiter radiates two to three times more heat than it receives from the Sun. This means that Jupiter is losing heat at a tremendously rapid rate.

The temperature at Jupiter's center may be about 30,000° C (54,000° F).

The best explanation for Jupiter's tremendous internal heat is that it is primordial heat left over from the heat of formation of the planet at the time the solar system was formed. An alternative explanation by some scientists for part of Jupiter's internal heat is that it is energy released by the fractionation of hydrogen and helium, a process believed to be currently under way somewhere near the center of the planet.

The displacement of the center of Jupiter's magnetic field from the center of the planet, and tilt of the field further confirm that Jupiter is a huge, flattened, fast-spinning ball of liquid hydrogen.

Jupiter's exact mass is 317.8 times Earth's mass, one lunar mass heavier than had been thought.

The rapid convection of heat in Jupiter's liquid interior is reflected in the constant rise and fall of the atmosphere. This is shown by a number of prominent, semi-permanent features seen in the planet's clouds, such as striking white ovals of rising atmosphere surrounded by darker borders of descending atmosphere.

Jupiter's atmosphere, the planet's 1,000-kilometer (600-mile)-deep outer layer, consists primarily of hydrogen and helium gas, with very small amounts of the other elements. *Pioneer 10* found that the atmosphere accounts for about 1 percent of Jupiter's total mass. Calculations based on *Pioneer 10* findings put the ratio of hydrogen to helium in Jupiter's upper atmosphere close to 80% to 20%, with less than 1% for all of the other elements. This is similar to the ratio of elements found in the Sun.

Jupiter's 17 relatively permanent belts and zones appear to be comparable to the continent-spanning cyclones and anticyclones which produce most of the weather in the Earth's temperate zones. On both planets these phenomena are huge regions of rising or falling atmospheric gas, powered by the Sun (and in Jupiter's case also by its internal heat source). On Earth, huge masses of warm, light gas rise to high altitudes, cool off, get heavier, and then roll down the sides of new rising columns of gas. The general direction of this atmospheric heat flow on Earth is from the tropics toward the poles. Coriolis forces (produced by planetary rotation) cause the descending gas, which would normally move north or south, to flow around the planet west to east. However, on Earth, unstable flow converts this west-east motion into the enormous spirals known as cyclones and anticyclones.

With Jupiter's high-speed, 22,000 mph rotation, these round-the-planet coriolis forces are very strong. Because of instabilities like those on Earth, flow should be in even more violent spirals than Earth's. But several factors appear to have a "calming effect," so that the motion is mostly linear. These factors include Jupiter's internal heat source and heat circulation, lack of a solid surface and liquid character. As a result, the combination of convection due to the Sun's heat plus internal heat, and coriolis due to Jupiter's rapid rotation, stretches the planet's large permanent weather features completely around the planet, forming the belts and zones.

While *Pioneer 10* did not completely explain Jupiter's Great Red Spot, it appears to be a centuries-old, 40,000-kilometer (25,000-mile)-wide vortex of a violent storm, as first proposed by the late Dr. Gerard P. Kuiper. The Spot is calculated to rise some 8 kilometers (5 miles) above the surrounding cloud deck. This is shown by the fact that the clouds at the top of the Spot have less atmosphere above them and are cooler (and hence higher) than surrounding clouds.

Pioneer 11 – Man's second spacecraft to Jupiter, *Pioneer 11*, reached the giant planet early in December 1974, almost exactly one year after *Pioneer 10*, and nearly two years after launch. *Pioneer 11* slid past Jupiter at a distance of just 41,000 kilometers (26,600 miles) – three times closer than *Pioneer 10* – experiencing intense radiation never before encountered by any spacecraft.

Pioneer 11 was the first spacecraft to make use of gravity assist – using the gravity of Jupiter to add to its velocity and change its direction – to fly on to its next destination, Saturn. Gravity assist has since become an essential maneuver for continued exploration of the outer solar system. *Pioneer 11* swept past Jupiter at more than 171,000 kilometers (107,000 miles) an hour, setting a record for the highest speed ever achieved by a manmade object. The spacecraft's trajectory took it around the planet against Jupiter's direction of rotation, permitting for the first time a look at a complete revolution of the planet's magnetic field, radiation belt and surface.

Like its predecessor, *Pioneer 11* carried a plaque telling any intelligent species which might find it several million years from now, who sent it and where it came from.

Voyager 1 and Voyager 2 –Following after the *Pioneer 10* and *11* missions were *Voyager 1* and *Voyager 2*, both launched in the late summer of 1977. Building on the results of the *Pioneers*, the Voyager missions extended our knowledge of Jupiter. Mission planning placed importance on confirming and clarifying the measurements made by the *Pioneers*. *Voyager 1*'s closest approach to Jupiter occurred March 5, 1979. *Voyager 2*'s closest approach was July 9, 1979.

Voyager photography of Jupiter actually began in January 1979, when images of the brightly banded planet already exceeded the best taken from Earth. *Voyager 1* completed its Jupiter encounter in early April, after taking almost 19,000 pictures and many other scientific measurements. *Voyager 2* picked up the baton in late April and its encounter continued into August. They took more than 33,000 pictures of Jupiter and its five major satellites.

Although astronomers had studied Jupiter from Earth for several centuries, scientists were surprised by many of the Voyager findings. Important physical, geological, and atmospheric processes were observed – in the planet, its satellites, and magnetosphere –that were new to observers.

Discovery of active volcanism on the satellite Io was probably the greatest surprise. It was the first time active volcanoes had been seen on another body in the solar system. It appears that activity on Io affects the entire Jovian system. Io appears to be the primary source of matter that pervades the Jovian magnetosphere — the region of space that surrounds the planet, primarily influenced by the planet's strong magnetic field. Sulfur, oxygen, and sodium, apparently erupted by Io's volcanoes and sputtered off the surface by impact of high-energy particles, were detected at the outer edge of the magnetosphere.

Galileo — More than a decade elapsed until the next Jupiter mission, Galileo, was launched in 1989. *Galileo* arrived at Jupiter in December 1995, after a six year journey. *Galileo* was the first spacecraft to actually orbit Jupiter (or any outer solar system planet). The *Galileo* spacecraft released a descent probe into the Jovian atmosphere. For a brief period, until it succumbed to Jupiter's immense atmospheric pressure, this probe provided our first direct measurements of temperature, pressure and composition of the Jovian atmosphere.

The *Galileo* spacecraft had two main sections, joined together by a spin bearing somewhat like a lazy susan. Half of the spacecraft contained pointable instruments, such as cameras, and maintained a fixed orientation relative to surrounding space. The other half of the spacecraft contained the instruments that measured magnetic fields and charged particles, and it slowly rotated in order to optimize these measurements.

Galileo was carried into Earth orbit in the cargo bay of the Space Shuttle *Atlantis*. It was then propelled onto its interplanetary flight path by a two-stage solid-fuel motor called an Inertial Upper Stage.

En route to Jupiter, *Galileo* flew close to two asteroids, the first-ever spacecraft to do so. It encountered the asteroid

Gaspra on October 29, 1991, and the asteroid Ida on August 28, 1993. During the latter part of its interplanetary cruise, *Galileo* was used to observe the collisions of fragments of Comet Shoemaker-Levy with Jupiter in July 1994. *Galileo* arrived at Jupiter on December 7, 1995, entering orbit and dropping its instrumented probe into the giant planet's atmosphere, and then began orbiting Jupiter.

Galileo discovered strong evidence that Jupiter's moon Europa has a melted salt water ocean under an ice layer on its surface. The spacecraft also found indications that two other moons, Ganymede and Callisto, have layers of liquid salt water. Other major science results from the mission include details of varied and extensive volcanic processes on the moon Io, measurements of conditions within Jupiter's atmosphere, and discovery of a magnetic field generated by Ganymede.

The mission ended on September 21, 2003, when the spacecraft plunged into Jupiter's atmosphere. This planned maneuver prevented the risk of Galileo drifting to an unwanted impact with the moon Europa, which may harbor a subsurface ocean.

Ulysses –A joint project between NASA and the European Space Agency, *Ulysses* was the first spacecraft sent out of the ecliptic (the plane in which Earth and other planets orbit the Sun) to study the Sun's north and south poles. The prime mission concluded in 1995, but *Ulysses* continued to monitor the Sun.

The Space Shuttle *Discovery* launched the *Ulysses* spacecraft in October 1990. To reach high solar latitudes, the spacecraft was aimed close to Jupiter so that Jupiter's large gravitational field would accelerate *Ulysses* out of the ecliptic plane to high latitudes.

Although the Ulysses program did not study deep space, it qualifies as a deep space mission by having flown out to Jupiter, twice, to perform gravity assist maneuvers.

Cassini-Huygens – The Cassini mission performed a Jupiter flyby in December 2000 on its way to Saturn. Although it was not long in Jupiter's vicinity, *Cassini* did acquire photographs and data using cameras and instrumentation more advanced than on any previous flight. (See also the Saturn section of this guide.)

Juno – The NASA Juno mission has been proposed for launch in 2010, with the goal of conducting an in-depth study of Jupiter. Looking deep into Jupiter's atmosphere, the mission should reveal fundamental processes of the formation and early evolution of our solar system. Juno's aim is to understand the origin and evolution of Jupiter, which will lead to a better understanding of our solar system and other planetary systems.

The Juno spacecraft will make maps of the gravity, magnetic fields, and atmospheric composition of Jupiter while in polar orbit around the planet (to date, all observations of Jupiter have been from more or less equatorial orbits).

SATURN

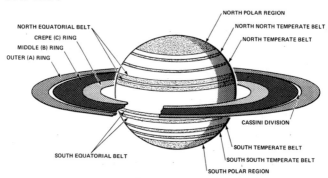

The visible surface of Saturn has a banded cloud structure much like that of Jupiter.

JUPITER – Mission Summaries

Pioneer 10:
  Launch: March 2, 1972
  Jupiter Flyby: December 3, 1973
  Status: Mission Ended

Pioneer 11:
  Launch: April 5, 1973
  Jupiter Encounter: February 2, 1974
  Status: Mission Ended

Voyager 2:
  Launch: August 20, 1977
  Jupiter Flyby: July 9, 1979
  Status: Headed to Interstellar Space

Voyager 1:
  Launch: September 5, 1977
  Jupiter Flyby: March 5, 1979
  Status: Headed to Interstellar Space

Galileo:
  Launch: October 18, 1989
  Jupiter Orbit Insertion: December 8, 1995
  Jupiter Impact: September 21, 2003
  Status: Mission Ended

Ulysses:
  10.06.90 Launch: October 6, 1990
  1st Jupiter Gravity Assist: February 8, 1991
  2nd Jupiter Gravity Assist: February 4, 2004
  Status: Returning to the Sun

Cassini-Huygens:
  Launch: October 15, 1997
  Jupiter Flyby: December 30, 2000
  Jupiter Observations Complete: March 31, 2001
  Status: Orbiting Saturn

Juno:
  Launch: 2010 (proposed)

SATURN – Facts & Figures
The orbit of Saturn is 8.54 AU from Earth.   Saturn has 46 known moons.
Average Distance from the Sun: 1,426,725,400 km / 885,904,700 miles / 9.53707 AU
Perihelion (closest approach to the Sun):   1,349,467,000 km / 838,519,000 miles / 9.021 AU
Aphelion (farthest distance from the Sun):   1,503,983,000 km / 934,530,000 miles / 10.054 AU
Equatorial Radius: 60,268 km / 37,449 miles (9.449 x radius of Earth)
Equatorial Circumference: 378,675 km / 235,298 miles
Volume: 8.271×10^{14} km^3 / 1.985×10^{14} miles^3 (763.6 x Earth's volume)
Mass: 5.685×10^{26} kg (95.16 x Earth's mass)
Density: 0.70 grams/cm^3 (0.127 x Earth's density)
Surface Area: 4.347×10^{10} km^2 / 1.678×10^{10} miles^2 (85.2 x Earth's surface)
Equatorial Surface Gravity: 7.207 m/sec^2 / 23.64 ft/sec^2 (0.74 x Earth weight)
Escape Velocity: 127,760 km/h / 79,390 mph (3.17 x Earth's escape velocity)
Sidereal Rotation Period (Length of Day):   0.44401 Earth days / 10.656 hours
Sidereal Orbit Period (Length of Year):   29.4 Earth years / 10755.7 Earth days
Mean Orbital Velocity: 34,821 km/h / 21,637 mph (0.865 x Earth orbital velocity)
Orbital Eccentricity: 0.0541506 (3.24 x Earth's orbital eccentricity)
Orbital Inclination to Ecliptic: 2.484 degrees

SATURN – Facts & Figures
Orbital Inclination to Ecliptic: 2.484 degrees
Equatorial Inclination to Orbit: 26.73 degrees (1.14 x Earth's equatorial inclination)
Orbital Circumference: 8,725,000,000 km / 5,421,000,000 miles (9.439 x Earth)
Effective Temperature: -178 °C / -288 °F / 95° K

Pioneer 11 – *Pioneer 11* exceeded the achievements of *Pioneer 10* by continuing on (with a gravity assist from Jupiter) to Saturn – the first manmade object to reach the ringed planet 2.4 billion kilometers (1.5 billion miles) from Earth orbit.

In September 1979 *Pioneer 11* flew to within 13,000 miles of Saturn and took the first close-up pictures of the planet. Instruments located two previously undiscovered small moons and an additional ring, charted Saturn's magnetosphere and magnetic field and found that its planet-size moon, Titan, was too cold for life.

Speeding underneath Saturn's ring plane, *Pioneer 11* sent back amazing pictures of the rings, which appeared dark, while the gaps between the rings appeared bright (the opposite of what we see from the Earth, because they were backlighted by the Sun in this instance).

Following its encounter with Saturn, *Pioneer 11* explored the outer regions of our Solar system, studying energetic particles from our Sun (Solar Wind) and cosmic rays entering our portion of the Milky Way. By September 1995, *Pioneer 11* could no longer make any scientific observations, and so routine daily mission operations were stopped.

Voyager 1 and Voyager 2 – The *Voyager 1* and *2* Saturn encounters occurred nine months apart, in November 1980 and August 1981, after which *Voyager 1* left the solar system and *Voyager 2* headed for and encounter with Uranus. The close-range observations of Saturn made by *Voyager 1* and *2* provided high-resolution data far different from the picture assembled during centuries of Earth-based studies.

The Voyagers found that Saturn's atmosphere is almost entirely hydrogen and helium. *Voyager 1* found that about 7% of the volume of Saturn's upper atmosphere is helium (compared with 11% of Jupiter's atmosphere), while almost all the rest is hydrogen. Also confirmed was Saturn's low density — it is the only planet in the solar system that is less dense than water.

While *Voyager 1* saw few markings, *Voyager 2*'s more sensitive cameras saw many — long-lived ovals, tilted features in east-west shear zones, and others similar to (but generally smaller than) those on Jupiter. Near Saturn's equator, the *Voyagers* measured winds of about 500 meters per second (1,100 mph), blowing mostly in an easterly direction. The dominance of eastward jet streams indicates that winds are not confined to the cloud layer, but must extend inward at least 2,000 kilometers (1,200 miles). Both Voyagers measured the rotation of Saturn (the length of a day) at 10 hours, 39 minutes and 24 seconds.

Cassini — After a flight of nearly seven years, which included four gravity assist maneuvers, the *Cassini* spacecraft — the largest and most massive spacecraft ever launched by the U.S. — went into orbit around Saturn in 2004. *Cassini* carried with it the European Space Agency's *Huygens* probe, which it released on an intersect course with Saturn's largest moon, Titan. *Huygens* entered the atmosphere of Titan and parachuted to the surface on January 14, 2005. It radioed scientific data to *Cassini* during its descent and for a short period from Titan's surface. By design, *Huygens*' batteries would be depleted after a maximum of half an hour on the surface. *Huygens* appears to have performed exactly as intended.

The *Cassini* spacecraft has been providing information about Saturn far in excess of what was learned during the Voyager program. For example, the spacecraft's Visual and Infrared Mapping Spectrometer, which sees deeper into the atmosphere than visible light, shows many more bands at lower altitudes than can be seen higher up, and these deeper bands show much more turbulent structure. Also, lightning appears to exist on Saturn. Instruments on the *Cassini* spacecraft have detected large radio emissions that seem to correspond with large storms visible in the atmosphere. Other new discoveries include wandering and rubble-pile moons; new and clumpy Saturn rings; splintering storms and a dynamic magnetosphere.

Another discovery was a tiny moon, about 5 kilometers (3 miles) across, recently named Polydeuces. Polydeuces is a companion, or "Trojan" moon of Dione (occupying the same orbit as Dione, but 60° degrees away). Saturn is the only planet known to have moons with Trojan companions.

Several faint Saturn rings have been discovered in Cassini images. Some lie in various gaps in the rings and may indicate the presence of tiny embedded moons acting as shepherds. Several of the rings are kinked, likely evidence of nearby moons.

Scientists also found that Saturn's winds change with altitude, and small storms emerge out of large ones. For the first time, *Cassini* images captured possible evidence of processes that may maintain the winds on Saturn. The observations offer a glimpse into the process which transfers energy by convection from Saturn's interior to help sustain strong winds.

SATURN – Mission Summaries

Pioneer 11:
  Launch: April 5, 1973
  Saturn Flyby: September 1, 1979
  Status: Mission Ended

Voyager 2:
  Launch: August 20, 1977
  Saturn Flyby: August 26, 1981
  Status: Headed to Interstellar Space

Voyager 1:
  Launch: September 5, 1977
  Saturn Flyby: November 12, 1980
  Status: Headed to Interstellar Space

Cassini-Huygens:
  Launch: October 15, 1997
  Phoebe Flyby: June 11, 2004
  Saturn Orbit Insertion: July 1, 2004
  Huygens Probe Release: December 24, 2004
  Huygens Probe Descent: January 14, 2005
  Status: Orbiting Saturn

URANUS

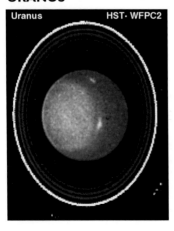

This NASA Hubble Space Telescope image of the planet Uranus reveals the planet's rings and bright clouds and a high altitude haze above the planet's south pole.

URANUS – Facts & Figures
The orbit of Uranus is 18.218 AU from Earth. Uranus was discovered By William Herschel, in 1781, and has 27 known moons.
Average Distance from the Sun: 2,870,972,200 km / 1,783,939,400 miles / 19.191 AU
Perihelion (closest approach to the Sun): 2,735,560,000 km / 1,699,800,000 miles / 18.286 AU
Aphelion (farthest distance from the Sun): 3,006,390,000 km / 1,868,080,000 miles / 20.096 AU
Equatorial Radius: 25,559 km / 15,882 miles (4.007 x radius of Earth)
Equatorial Circumference: 160,592 km / 99,787 miles
Volume: 6.914×10^{13} km^3 / 1.659×10^{13} miles^3 (63.1 x Earth's volume)
Mass: 8.685×10^{25} kg (14.371 x Earth's mass)
Density: 1.30 grams/cm^3 (0.236 x Earth's density)
Surface Area: 8.116×10^9 km^2 / 3.133×10^9 miles^2 (15.91 x Earth's surface)
Equatorial Surface Gravity: 8.43 m/sec^2 / 27.7 ft/sec^2 / (0.86 x Earth weight)
Escape Velocity: 76,640 km/h / 47,620 mph (1.90 x Earth's escape velocity)
Sidereal Rotation Period (Length of Day): -0.7196 Earth days (retrograde) / -17.24 hours (retrograde)
Sidereal Orbit Period (Length of Year): 84.02 Earth years / 30,687.2 Earth days
Mean Orbital Velocity: 24,607 km/h / 15,290 mph (0.229 x Earth orbital velocity)
Orbital Eccentricity: 0.047168 (2.823 x Earth's orbital eccentricity)
Orbital Inclination to Ecliptic: 0.770 degrees

URANUS – Facts & Figures
Equatorial Inclination to Orbit: 97.86 degrees (4.173 x Earth's equatorial inclination)
Orbital Circumference: 17,620,000,000 km / 10,948,560,400 miles (19.06 x Earth)
Effective Temperature: -216 °C / -357 °F / 57° K

Voyager 2 –Voyager 2 greatly extended man's conquest of the solar system by continuing on past Jupiter and Saturn to fly by both Uranus and Neptune, the first and only spacecraft to do so. This four-planet mission, referred to as the Grand Tour, took advantage of a rare alignment of the planets whereby the spacecraft's velocity allowed it to intersect each planet's orbit at a time when the planet was at the same location. This planetary alignment repeats every 176 years.

The *Voyager 2* spacecraft flew closely past distant Uranus, the seventh planet from the Sun, in January 1986. At its closest, the spacecraft came within 81,500 kilometers (50,600 miles) of Uranus's cloud tops. *Voyager 2* radioed thousands of images and voluminous amounts of other scientific data on the planet, its moons, rings, atmosphere, interior and the magnetic environment surrounding Uranus.

Voyager 2's images of the five largest moons around Uranus revealed complex surfaces indicative of varying geologic pasts. The cameras also detected 10 previously unseen moons. Several instruments studied the ring system, uncovering the fine detail of the previously known rings and two newly detected rings. The spacecraft also found a Uranian magnetic field that is both large and unusual. In addition, the temperature of the equatorial region, which receives less sunlight over a Uranian year, is nevertheless about the same as that at the poles.

In December 2005, NASA's Hubble Space Telescope photographed a new pair of rings around Uranus and two new, small moons orbiting the planet. The largest ring is twice the diameter of the planet's previously known rings. The rings are so far from the planet that they are being called Uranus' "second ring system." One of the new moons shares its orbit with one of the rings. Analysis of the Hubble data also reveals the orbits of Uranus' family of inner moons have changed significantly over the past decade.

URANUS – Mission Summaries

Voyager 2:
  Launch: August 20, 1977
  Uranus Flyby: January 24, 1986
  Status: Headed to Interstellar Space

NEPTUNE

NEPTUNE – Facts & Figures
The orbit of Neptune is 29.060 AU from Earth. Neptune was discovered By Johann Galle, in 1846, and has 13 known moons.
Average Distance from the Sun: 4,498,252,900 km / 2,795,084,800 miles / 30.069 AU
Perihelion (closest approach to the Sun):   4,459,630,000 km / 2,771,087,000 miles / 29.811 AU
Aphelion (farthest distance from the Sun):   4,536,870,000 km / 2,819,080,000 miles / 30.327 AU
Equatorial Radius: 24,764 km / 15,388 miles (3.883 x radius of Earth)
Equatorial Circumference: 155,597 km / 96,683 miles
Volume: 6.253×10^{13} km^3 / 1.500×10^{13} miles^3 (57.7 x Earth's volume)
Mass: 1.024×10^{26} kg (17.147 x Earth's mass)

NEPTUNE – Facts & Figures
Density: 1.76 grams/cm^3 (0.317 x Earth's density)
Surface Area: 7.641 x 10^9 km^2 / 2.950 x 10^9 miles^2 (14.980 x Earth's surface)
Equatorial Surface Gravity: 10.71 m/sec^2 / 35.14 ft/sec^2 / (1.1 x Earth weight)
Escape Velocity: 85,356 km/h / 53,038 mph (2.12 x Earth's escape velocity)
Sidereal Rotation Period (Length of Day):   0.67125 Earth days / 16.11 hours
Sidereal Orbit Period (Length of Year):   164.79 Earth years / 60,190 Earth days
Mean Orbital Velocity: 19,720 km/h / 12,253 mph (0.490 x Earth orbital velocity)
Orbital Eccentricity: 0.00859 (0.514 x Earth's orbital eccentricity)
Orbital Inclination to Ecliptic: 1.769 degrees
Equatorial Inclination to Orbit: 29.58 degrees (1.261 x Earth's equatorial inclination)
Orbital Circumference: 28,142,000,000 km / 17,487,000,000 miles (30.44 x Earth)
Effective Temperature: -214 °C / -353 °F / 59° K

Pioneer 11 – *Pioneer 11* crossed the orbit of Neptune in 1990 and become the fourth spacecraft to leave the solar system. When it crossed Neptune's orbit, *Pioneer 11* was 2.8 billion miles from the Earth. Neptune's orbit marks one measure of the expanse of the solar system because Pluto's highly eccentric orbit carries it inside the path of Neptune's orbit for years at a time.

Voyager 2 – In the summer of 1989, *Voyager 2* became the first spacecraft to observe the planet Neptune. Passing about

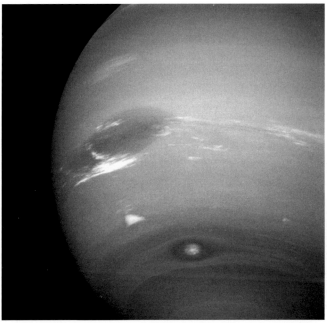

This photograph of Neptune shows three of the features that Voyager 2 photographed. At the north is the Great Dark Spot, accompanied by bright, white clouds that undergo rapid changes in appearance. To the south of the Great Dark Spot is the bright feature that Voyager scientists have nicknamed 'Scooter." Still farther south is the feature called 'Dark Spot 2," which has a bright core. Each feature moves eastward at a different velocity, so it is only occasionally that they appear close to each other, such as at the time this picture was taken.

4,950 kilometers (3,000 miles) above Neptune's north pole, *Voyager 2* made its closest approach to any planet. Five hours later, Voyager 2 passed about 40,000 kilometers (25,000 miles) from Neptune's largest moon, Triton, the last solid body the spacecraft had an opportunity to study.

Voyager 2 traveled 12 years at an average velocity of 19 kilometers a second (about 42,000 miles an hour) to reach Neptune, which is 30 times farther from the Sun than Earth is. *Voyager 2* observed Neptune almost continuously from June to October 1989. *Voyager 2* is now headed out of the solar system.

NEPTUNE – Mission Summaries
Voyager 2: 　Launch: August 20, 1977 　Neptune Flyby: August 24-25, 1989 　Status: Headed to Interstellar Space

PLUTO

New Horizons – Pluto, the smallest planet, is the only planet not yet visited by a spacecraft. Most of what we know about Pluto has been learned since the late 1970s from ground-based observations, the Infrared Astronomical Satellite (IRAS), and the Hubble Space Telescope. Many of the key questions about Pluto and its satellite Charon await the close-up observation of a space flight mission.

In November 2001, NASA selected a mission for preliminary design studies. This mission, *New Horizons: Shedding Light on Frontier Worlds*, proposes a spacecraft which would use a remote sensing package (that includes imaging instruments and a radio science investigation) as well as spectroscopic and other experiments, to characterize the global geology and morphology of Pluto and its moon Charon. It will also map their surface compositions and characterize Pluto's neutral atmosphere and its escape rate. Scientists expect to find a complex world – with areas "darker than coal" and "brighter than snow." The mission plan also calls for a fly by of Kuiper Belt Objects by 2022.

PLUTO – Facts & Figures
The (average) orbit of Pluto is 38.5 AU from Earth. Pluto was discovered By Clyde Tombaugh, in 1930, and has 1 (3?) known moons.
Average Distance from the Sun: 5,906,380,000 km / 3,670,050,000 miles / 39.482 AU
Perihelion (closest approach to the Sun): 4,436,820,000 km / 2,756,902,000 miles / 29.658 AU
Aphelion (farthest distance from the Sun): 7,375,930,000 km / 4,583,190,000 miles / 49.305 AU
Equatorial Radius: 1,151 km / 715 miles (0.180 × radius of Earth)
Equatorial Circumference: 7,232 km / 4,494 miles
Volume: 6.390×10^9 km^3 / 1.530×10^9 miles^3 (0.0059 × Earth's volume)
Mass: 1.300×10^{22} kg (0.0022 × Earth's mass)
Density: 2 grams/cm^3 (approx. 0.4 × Earth's density)
Surface Area: 1.665×10^7 km^2 / 6.430×10^6 miles^2 (0.033 × Earth's surface)
Equatorial Surface Gravity: 0.81 m/sec^2 / 2.7 ft/sec^2 / (0.08 × Earth weight)
Escape Velocity: 4,570 km/h / 2,840 mph (0.11 × Earth's escape velocity)
Sidereal Rotation Period (Length of Day): 6.387 Earth days / 153.3 hours
Sidereal Orbit Period (Length of Year): 247.92 Earth years / 90,553 Earth days
Mean Orbital Velocity: 17,096 km/h / 10,623 mph (0.425 × Earth orbital velocity)
Orbital Eccentricity: 0.2488 (14.9 × Earth's orbital eccentricity)

PLUTO – Facts & Figures
Orbital Inclination to Ecliptic: 17.14 degrees
Equatorial Inclination to Orbit: 119.61 degrees (5.10 x Earth's equatorial inclination)
Orbital Circumference: 32,820,000,000 km / 20,390,000,000 miles (35.505 x Earth)
Minimum – Maximum Surface Temperature: -233 – -223 °C / -387 – -369 °F / 40 – 50 °K

New moons –In November 2005, NASA announced that, using Hubble Space Telescope, astronomers discovered that Pluto may have not one, but three moons. The candidate moons, provisionally designated S/2005 P1 and S/2005 P2, are approximately 27,000 miles (44,000 kilometers) away from Pluto –two to three times as far from Pluto as Charon. The estimated diameters of these tiny moons lie between 40 and 125 miles (64 and 200 kilometers). Charon, for comparison, is about 730 miles (1170 km) wide, while Pluto itself has a diameter of about 1,410 miles (2270 km). Follow-up Hubble observations are planned for February 2006 to confirm that the newly discovered objects are truly Pluto's moons. Only after confirmation will the International Astronomical Union consider permanent (and catchier) names for S/2005 P1 and S/2005 P2.

PLUTO – Mission Summaries

New Horizons (PKB):
  Launch: Janusry 2006
  Pluto Arrival Window: November 17, 2016 - July 11, 2017
  Flyby of Kuiper Belt Objects: 2018 - 2022
  Status: Final preparations and testing

Part 3 – The Milky Way Galaxy
DETECTING EXTRASOLAR PLANETS

Keck Interferometer – The Keck Interferometer is part of NASA's overall effort to find planets – and ultimately life – beyond our solar system. It will combine the light from the twin Keck telescopes to: 1) measure the emission from dust orbiting nearby stars; 2) directly detect the hottest gas giant planets; 3) image disks around young stars and other objects of astrophysical interest; and 4) survey hundreds of stars for the presence of planets the size of Uranus or larger. At 4,150 meters (13,600 feet) above the Pacific Ocean, atop the dormant volcano Mauna Kea on the 'Big Island" of Hawaii, the twin Keck Telescopes are the world's largest telescopes for optical and near-infrared astronomy. The Keck Interferometer joins these giant telescopes to form a powerful astronomical instrument.

LBTI – The Large Binocular Telescope Interferometer (LBTI) will study the formation of solar systems and will be capable of directly detecting giant planets outside our solar system. Two 8-meter class telescopes on Mount Graham, Arizona, will be connected in an infrared interferometer. Because of its unique geometry and relatively direct optical path, the LBTI will offer science capabilities that are different from other interferometers. It will provide high-resolution images of many faint objects over a wide field-of-view, including galaxies in the 'Hubble Deep Field," but with 10 times the Hubble resolution. By studying and characterizing the emissions from faint dust clouds around other stars, LBTI will provide critically needed data for the design of the Terrestrial Planet Finder, a future mission that will study planets orbiting nearby stars.

Corot – is an ESA mission, the first mission with the ability to detect rocky planets (several times larger than Earth), orbiting nearby stars. *Corot* is a 30-centimetre space telescope and it is scheduled to be launched in early 2006. *Corot* will attempt to detect planets crossing in front a star by telescopically monitoring the changes in the star's brightness (the "transit" method). *Corot* will also be able to detect "starquakes" (ripples across a star's surface) the nature of which will provide astronomers with calculations for the star's mass, age and chemistry (a technique called asteroseismology). The *Corot* data will also be compared with equivalent data for the Sun.

Kepler – Kepler is a NASA spaceborne telescope designed to look for Earth-like planets around other stars. Detection will be done indirectly, using the "transit" method, as with *Corot* above. A transit occurs each time a planet crosses the line-of-sight between its star and the observer, blocking some of the light from the star. This periodic dimming is used to detect the planet and to determine its size and its orbit. Scheduled to launch in 2008, Kepler will hunt for planets using a specialized one-meter diameter telescope called a photometer to measure the small changes in brightness caused by transits.

SIM PlanetQuest – SIM will be an optical interferometer operating in an Earth-trailing solar orbit. SIM observations will be applied to multiple investigations related to planet detection and stellar astrophysics. Check-out and calibration of the interferometer in its 95-million-kilometer orbit, will continue for several months. For at least five years after the end of this calibration period the SIM interferometer will then perform nearly continuous science observations over the entire celestial sphere. Observation data will be stored onboard, and returned to Earth several times each week. Special "quick turnaround" procedures will be used to handle "targets of opportunity."

Terrestrial Planet Finder (TPF) – TPF is a suite of two complementary observatories that will study all aspects of planets outside our solar system: from their formation and development in disks of dust and gas around newly forming stars to the presence and features of those planets orbiting the nearest stars; from the numbers at various sizes and places to their suitability as an abode for life. By combining the high sensitivity of space telescopes with revolutionary imaging technologies, the TPF observatories will measure the size, temperature, and placement of planets as small as the Earth in the habitable zones of distant solar systems. In addition, TPF's spectroscopy will allow atmospheric chemists and biologists to use the relative amounts of gases like carbon dioxide, water vapor, ozone and methane to find whether a planet someday could or even now does support life.

Darwin – Darwin, an ESA program, will utilize four 'free-flying" spacecraft to search for Earth-like planets around other stars. It will also analyze their atmospheres for chemical signatures suggesting life. and perform high-resolution imaging to provide pictures of celestial objects with unprecedented detail. Darwin will use wavelengths in the mid-infrared where the contrast between a star and its planets is on the order of a million to one, instead of the billion to one brightness difference that is typical of optical wavelengths,

STUDYING STARS

Gaia – Gaia is an ESA mission that will chart the distances, movements, and changes in brightness of one billion stars in our Galaxy over a five-year period. Hundreds of thousands of new celestial objects are expected to be discovered. Gaia will also detect and characterize many thousands of extrasolar planetary systems in millions of galaxies.

SWAS – SWAS is one of NASA's Small Explorer Program (SMEX) missions. The overall goal of the mission is to gain a greater understanding of star formation by determining the composition of interstellar clouds and establishing the means by which these clouds cool as they collapse to form stars and planets.

Outrigger Telescopes Project (OTP) – The Outrigger Telescopes Project would combine the light of multiple telescopes using a technique called interferometry to search for planets around nearby stars, make images of newborn stars, and study faint, dim and distant objects beyond our galaxy. OTP will contribute to answering the questions "Where do we come from?" and "Are we alone?" by addressing four specific science objectives:

   Find planets around nearby stars.
   Look for newborn stars.
   View the faintest and farthest.
   See our solar system family up close.

Part 4 – The Universe

MICROWAVE

Wilkinson Microwave Anisotropy Probe (WMAP) – The WMAP team has made the first detailed full-sky map of the oldest light in the universe. It is a 'baby picture" of the universe. The microwave light captured in its images is from 379,000 years after the Big Bang, over 13 billion years ago. The data brings into high resolution the seeds that generated the cosmic structure we see today. These patterns are tiny temperature differences within an extraordinarily evenly dispersed microwave light bathing the universe, which now averages a frigid 2.73 degrees above absolute zero temperature. WMAP resolves slight temperature fluctuations, which vary by only millionths of a degree.

Planck –Planck is an ESA satellite that will look back to a time close to the Big Bang, and observe the oldest radiation in the Universe (the cosmic microwave background). The goal of Planck is to better understand how galaxies and galaxy clusters formed. Some of the key questions Planck will answer are:

   What is the age of the Universe?

Will the Universe continue its expansion forever, or will it collapse into a 'Big Crunch'?

What is the nature of the so-called 'dark matter' (which may account for more than 90% of the total amount of matter in the Universe but that has never been detected directly)?

INFRARED

Infrared Space Observatory (ISO) – ISO was an ESA program, completed in May 1998, which studied the dusty regions of the Universe, which are opaque to visible-light telescopes. ISO's infrared detectors provided a much more detailed look at the infrared view of universe than had been previously achieved, giving scientists a view of astronomical objects that are invisible at other wavelengths.

Spitzer – Launched in August 2003, the Spitzer Space Telescope, formerly known as the Space Infrared Telescope Facility (SIRTF), is an infrared telescope that will study the early universe, young galaxies and forming stars, and will detect dust discs around stars, considered an important signpost of planetary formation. An infrared cousin of Hubble, the Spitzer Space Telescope consists of a cryogenically cooled telescope with lightweight optics that deliver light to advanced, large-format infrared detector arrays. Spitzer orbits the Sun, trailing behind Earth, drifting in a benign thermal environment.

Herschel –The Herschel Space Observatory is a space-based telescope that will study the Universe by the light of the far-infrared and submillimeter portions of the spectrum. It is expected to reveal new information about the earliest, most distant stars and galaxies, as well as those closer to home in space and time. It is scheduled to be launched in 2007, and is expected to remain an active observatory for at least three years. Two-thirds of Herschel's observation time will be available to the world scientific community, with the remainder reserved for the spacecraft's science and instrument teams.

WISE —WISE is a NASA-funded unmanned satellite carrying an infrared-sensitive telescope that will study the solar system, the Milky Way, and the universe. Among the objects WISE will study are asteroids, the coolest and dimmest stars, and the most luminous galaxies. Orbiting several hundred miles above the dividing line between night and day on Earth, WISE will always point toward the Sun. After a period of six months WISE will have taken nearly 1,500,000 pictures and will have observed the entire sky.

James Webb Space Telescope (JWST) – JWST is a large, infrared-optimized space telescope scheduled for launch no earlier than June 2013. JWST is designed to study the earliest galaxies and some of the first stars formed after the Big Bang. These early objects have a high red shift from our vantage point, meaning that the best observations for these objects are available in the infrared. JWST's instruments also have some capability in the visible spectrum. JWST will reside in an L2 Sun-Earth Lagrange point Lissajous orbit, about 1.5 million km (1 million miles) from the Earth.

ULTRAVIOLET

International Ultraviolet Explorer (IUE) —IUE was an ESA satellite that performed ultraviolet spectroscopic observations of cosmic objects, from comets to quasars. The IUE program was completed in September 1996. IUE was the first astronomical satellite in high Earth orbit. It detected ultraviolet radiation that cannot penetrate Earth's ozone layer. Among its notable activities were: observations of Halley's Comet in 1986; the first space observations of a naked-eye-visible supernova event in 300 years; and an extensive observational program detailing the evolution of Jupiter's atmosphere after the impact of Comet Shoemaker-Levy in 1994.

Far Ultraviolet Spectroscopic Explorer (FUSE) —FUSE looks at light in the far ultraviolet portion of the electromagnetic spectrum (approximately 90 to 120 nanometers), which is unobservable with other telescopes. FUSE observes these wavelengths with much greater sensitivity and resolving power than previous instruments used to study light in this wavelength range. The FUSE science instrument collects the light of distant objects and contains the equipment necessary to disperse and record the light.

THE HUBBLE SPACE TELESCOPE –Hubble orbits 600 kilometers (375 miles) above Earth, viewing the universe in infrared, ultraviolet, and visible light. Its very precise pointing ability and powerful optics provide views of the universe that cannot be made using ground-based telescopes or other satellites. Hubble is the first scientific mission of any kind that is specifically designed for routine servicing by space walking astronauts. Every day, Hubble delivers between 10 and 15 gigabytes of data to astronomers all over the world. This has created a data archive of over 10 terabytes. More than 400,000 separate observations have been made.

X-RAYS

Exosat – Exosat was the first ESA mission to study the universe at X-ray wavelengths, detecting and observing high-energy sources. Completed in 1986, the Exosat program observed a wide variety of objects, including active galactic nuclei, X-ray binary systems, supernova remnants, and clusters of galaxies. Exosat was one of the first unmanned satellites to feature an on-board computer.

Rossi X-ray Timing Explorer (RXTE) – RXTE is a satellite that observes the fast-moving, high-energy worlds of black holes, neutron stars, X-ray pulsars and bursts of X-rays that light up the sky and then disappear forever. RXTE was launched into low-Earth orbit on December 30, 1995, and is still going strong, making unique contributions to the understanding of these extreme objects.

XMM-Newton —Launched in December 1999, the European Space Agency's X-ray Multi-Mirror satellite (XMM-Newton) is the most powerful X-ray telescope ever placed in orbit. It is the largest science satellite ever built in Europe. Earth's atmosphere blocks X-rays generated by violent processes such as black holes to the formation of galaxies. By placing X-ray detectors in space these sources can be detected and studied in detail. XMM-Newton, XMM-Newton carries three very advanced X-ray telescopes.

INTEGRAL —The INTErnational Gamma-Ray Astrophysics Laboratory (INTEGRAL) is an ESA mission in cooperation with Russia and the United States. It is the most sensitive gamma-ray observatory ever launched, and its original two-year mission has been extended until December 2010. Gamma-rays, even more powerful than X-rays, are shielded by the Earth's atmosphere so gamma-rays from space can only be detected by satellites. INTEGRAL captures data and images on violent explosions, the formation of elements, black holes and other exotic objects in the observable universe.

Nuclear Spectroscopic Telescope Array (NuSTAR) – NuSTAR will open the high-energy X-ray sky for sensitive study for the first time. X-ray telescopes like Chandra and XMM-Newton have peered deep into the X-ray universe at low X-ray energy. By focusing X-rays at higher energy; up to 80 keV, NuSTAR will answer fundamental questions about the Universe: How are black holes distributed through the cosmos? How were the elements that compose our bodies and the Earth forged in the explosions of massive stars? What powers the most extreme active galaxies? NuSTAR offers the opportunity to explore our Universe in an entirely new way.

X-ray Evolving Universe Spectroscopy (XEUS) – ESA's XEUS mission, currently in development, will provide a space-borne X-ray observatory (200 times more sensitive than

XMM-Newton and a potential follow-on to it) for studying black holes, galaxy groups, clusters and the interstellar medium. XEUS will consist of two spacecraft flying 50 meters apart —one carrying a mirror and one to carry the detectors.

GAMMA-RAY BURSTS

High Energy Transient Explorer (HETE-2) – HETE-2 is a small scientific satellite designed to detect and localize gamma-ray bursts (GRBs). The coordinates of GRBs detected by HETE are distributed to interested ground-based observers within seconds of burst detection, thereby allowing detailed observations of the initial phases of GRBs. The primary goal of HETE-2 is to determine the origin and nature of cosmic gamma-ray bursts (GRBs). This is accomplished through the simultaneous, broad-band observation in the soft X-ray, medium X-ray, and gamma-ray energy ranges, and the precise localization and identification of cosmic gamma-ray burst sources (GRBs).

Swift – NASA's Swift mission is a first-of-its-kind multi-wavelength observatory dedicated to the study of gamma-ray burst (GRB) science. Its three instruments will work together to observe GRBs and afterglows in the gamma-ray, X-ray, ultraviolet, and optical wavebands. The main mission objectives for Swift are to:

Determine the origin of gamma-ray bursts.
Classify gamma-ray bursts and search for new types.
Determine how the explosion develops.
Use gamma-ray bursts to study the early universe.
Perform the first sensitive hard X-ray survey of the sky.

During its mission, lasting at least two years, Swift is expected to observe more than 200 bursts. The Swift mission will represent the most comprehensive study of GRB afterglows to date.

STAR FORMATION

Hipparcos – ESA's Hipparcos mission, completed in 1993, mapped the heavens more accurately than ever before it, recording the stars, their motions and their distances. Hipparcos, observed more than 100,000 stars to high precision, and more than 1,000,000 stars with lesser precision. Each star that Hipparcos studied was observed about 100 times over a four-year period.

Galaxy Evolution Explorer (GALEX) – GALEX is an orbiting space telescope that makes observations at ultraviolet wavelengths to measure the history of star formation in the universe 80 percent of the way back to the Big Bang. Since scientists believe the universe is about 13 billion years old, the mission will study galaxies and stars across about 10 billion years of cosmic history. The spacecraft's mission is to observe hundreds of thousands of galaxies, with the goal of determining how far away each galaxy is from Earth and how fast stars are forming in each galaxy.

Single Aperture Far-Infrared Observatory (SAFIR) –SAFIR is a large cryogenic space telescope scheduled for launch in the 2015 to 2020 period. 'Single Aperture" refers to the telescope's single primary mirror, distinguishing it from multi-mirror interferometry missions. SAFIR will study the earliest phases of forming galaxies, stars, and planetary systems at wavelengths where these objects are brightest: from 20 microns to one millimeter. Most of this portion of the electromagnetic spectrum is not accessible from the ground because it is absorbed by moisture in Earth's atmosphere. Half or more of the optical and ultraviolet light produced in the universe is absorbed by dust and reradiated in the far-infrared and submillimeter ranges. Even in our local area of the universe, many galaxies are so dusty that they radiate mainly at those wavelengths.

A Boeing Delta II rocket hurls Deep Space 1 through the morning clouds after liftoff, creating sun-challenging light with its exhaust.

Photo of Comet Halley taken by the Giotto spacecraft in 1986.

Artist's concept of Deep Space 1 flying within 10 miles (15 kilometers) of asteroid 1992 KD.

A closeup view of the experimental solar-powered ion propulsion engine of Deep Space 1.

Deep Space Operations Center, also known as the Dark Room, is the heart of the Space Flight Operations Facility at JPL. JPL is the destination point for all the data beamed back to Earth via NASA's worldwide Deep Space Network from planetary and Earth-orbiting spacecraft, as well as spacecraft of other space agencies around the world.

In the clean room at Astrotech Space Operations near KSC, the plastic protective cover is lifted from the Deep Impact spacecraft prior to undergoing functional testing.

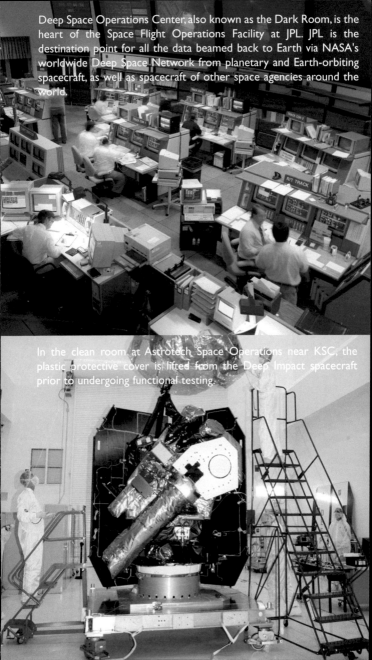

The sun rises behind Launch Pad 17-B, Cape Canaveral Air Force Station, Florida, where the Boeing Delta II rocket carrying the Deep Impact spacecraft waits for launch.

Artist's rendering of the Stardust spacecraft.

Composite image taken by Stardust's navigation camera during the close approach to Wild 2.

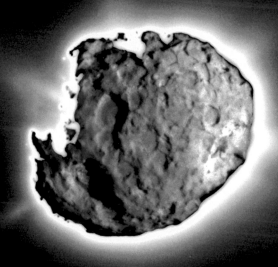

The Pioneer missions used an Atlas-Centaur booster with an added third stage to achieve a launch velocity of over 51,500 km (32,000 miles) per hour.

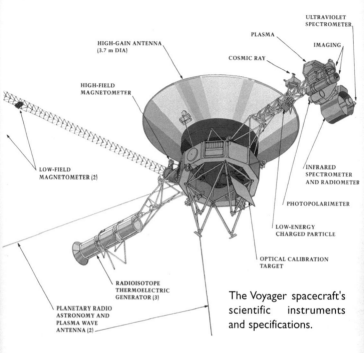

The Voyager spacecraft's scientific instruments and specifications.

VOYAGER SPACECRAFT FEATURES	
Spacecraft Weight	808 kg (1782 lb)
Science Instruments Weight	105 kg (232 lb)
High-Gain Antenna Diameter	3.7 m (12 ft)
Radioisotope Thermoelectric Generator (RTG) Power (at Saturn)	~400 W
Data Storage Capability	538 million bits
X-Band Data Rate at Jupiter at Saturn	115,200 bits per second 44,800 bits per second

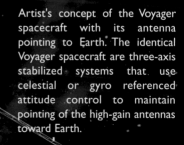

Artist's concept of the Voyager spacecraft with its antenna pointing to Earth. The identical Voyager spacecraft are three-axis stabilized systems that use celestial or gyro referenced attitude control to maintain pointing of the high-gain antennas toward Earth.

As Voyager 1 flew by Jupiter, it captured this photo of the Great Red Spot, an anti-cyclonic (high-pressure) storm on Jupiter that can be likened to the worst hurricanes on Earth.

Jupiter and its four planet-size moons, called the Galilean satellites, were photographed in early March 1979 by Voyager 1 and assembled into this collage.

Tracking of and communication with unmanned spacecraft are accomplished through the worldwide Deep Space Network. The main elements of NASA's Deep Space Network (DSN) are the Deep Space Stations based around the globe, the network control center, and the ground communications facility that links the stations and control center together.

The Voyager spacecraft weighs 795 kg (1,753 lb) including 113 kg (249 lb) of scientific instruments. The large antenna at the top is 3.66 m (12 ft) in diameter.

The Galileo spacecraft in the KSC Vertical Processing Facility being attached to the Inertial Upper Stage before mating with the launch vehicle.

The Galileo spacecraft and its Inertial Upper Stage being deployed in orbit by STS-34 Atlantis.

The launch of the Space Shuttle Atlantis in October 1989. The Shuttle cargo bay carried the Galileo spacecraft into Earth orbit.

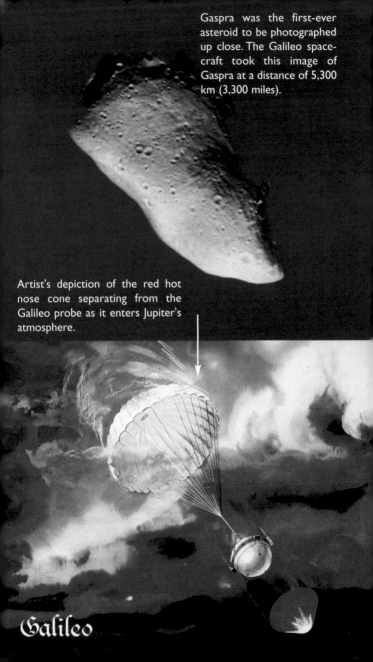

Gaspra was the first-ever asteroid to be photographed up close. The Galileo spacecraft took this image of Gaspra at a distance of 5,300 km (3,300 miles).

Artist's depiction of the red hot nose cone separating from the Galileo probe as it enters Jupiter's atmosphere.

Galileo

This true-color composite frame, made December 12, 2000, captures Io and its shadow in transit against the disk of Jupiter. The distance of the spacecraft from Jupiter was 19.5 million kilometers (12.1 million miles).

Galileo acquired its highest resolution images of Jupiter's moon Io on July 3, 1999 during its closest pass to Io since orbit insertion in late 1995.

Cutaway views of the possible internal structures of the Galilean satellites as determined by Voyager and Galileo data.

Io
Radius: 1821 km

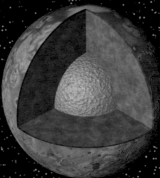

Io's rock or silicate shell extends to the surface, while the rock layers of Ganymede and Europa are in turn surrounded by shells of water in ice or liquid form (shown in blue and white).

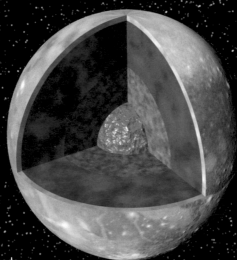

Ganymede
Radius: 2634 km

With the exception of Callisto, all the satellites have metallic (iron, nickel) cores (shown in gray). With the exception of Callisto, all the cores are surrounded by rock (shown in brown) shells.

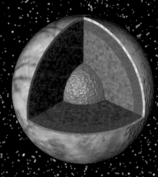

Europa
Radius: 1565 km

Callisto is shown as a relatively uniform mixture of comparable amounts of ice and rock. Recent data, however, suggests a more complex core as shown here (bottom right).

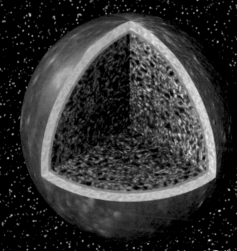

Callisto
Radius: 2403 km

The solar system's largest moon, Ganymede, is captured here alongside the planet Jupiter in a color picture taken by Cassini on December 3, 2000. Cassini was 26.5 million kilometers (16.5 million miles) from Ganymede when this image was taken.

This true-color simulated view of Jupiter is composed of 4 images taken by Cassini on December 7, 2000.

The Great Red Spot has been present in Jupiter's atmosphere for at least 300 years. It is a vast storm, spinning like a cyclone. This Hubble Space Telescope mosaic presents a series of pictures of the Red Spot obtained by Hubble between 1992 and 1999.

Hubble
Heritage

This is a spectacular Hubble Space Telescope close-up view of an electric-blue aurora that is eerily glowing one half billion miles away on the giant planet Jupiter. Auroras are curtains of light resulting from high-energy electrons racing along the planet's magnetic field into the upper atmosphere.

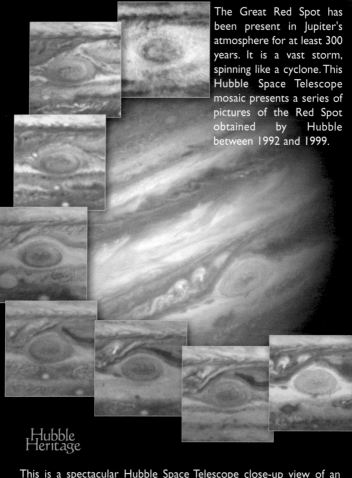

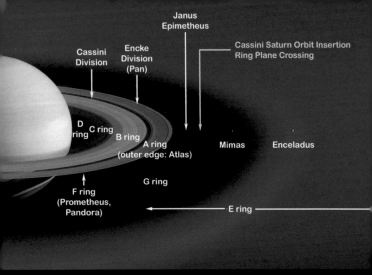

This montage of images of the Saturnian system was prepared from an assemblage of images taken by the Voyager 1 spacecraft during its Saturn encounter in November 1980. This artist's view shows Dione in the forefront, Saturn rising behind, Tethys and Mimas fading in the distance to the right, Enceladus and Rhea off Saturn's rings to the left, and Titan in its distant orbit at the top.

This is an artist's concept of Saturn's rings and major icy moons. Saturn's rings make up an enormous, complex structure. From edge-to-edge, the ring system would not even fit in the distance between Earth and the Moon. The seven main rings are labeled in the order in which they were discovered. From the planet outward, they are D, C, B, A, F, G and E.

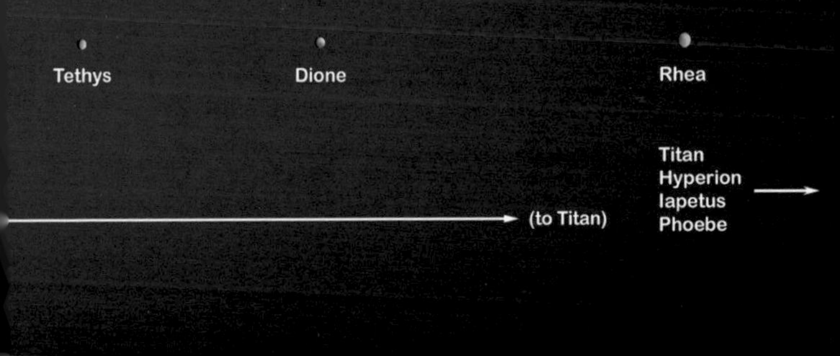

Tethys

Dione

Rhea

Titan
Hyperion
Iapetus
Phoebe

(to Titan) →

Artists concept of the Cassini Saturn Orbiter and Titan Probe Spacecraft.

Cassini's seven-year journey to Saturn began with the liftoff of a Titan IVB/Centaur carrying the orbiter and its attached Huygens probe. Launch occurred on October 15, 1997 from Launch Complex 40 on Cape Canaveral Air Station.

This Voyager 2 view, focusing on Saturn's C-ring was compiled from three separate images taken through ultraviolet, clear and green filters on August 23, 1981.

This artist's concept shows the European Space Agency's Huygens probe descent sequence. The animation shows the Huygens probe's entry, descent and landing, with the descent imager/spectral radiometer lamp turned on at the end.

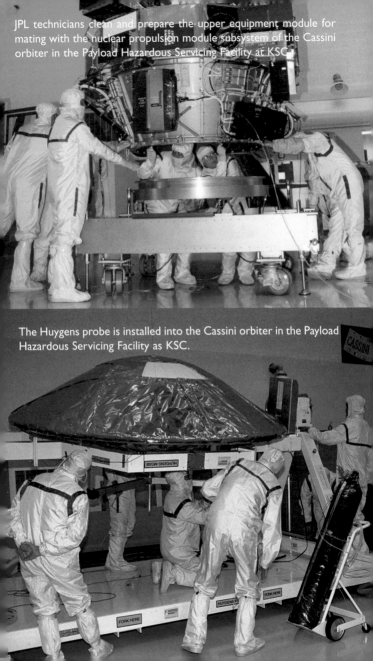

JPL technicians clean and prepare the upper equipment module for mating with the nuclear propulsion module subsystem of the Cassini orbiter in the Payload Hazardous Servicing Facility at KSC.

The Huygens probe is installed into the Cassini orbiter in the Payload Hazardous Servicing Facility as KSC.

The completed Cassini Spacecraft in a JPL Assembly Room. The orbiter mass at launch was nearly 5300 kg, over half of which is propellant for trajectory control. The mass of the Titan probe (2.7 m diameter) is roughly 350 kg.

As it swooped past the south pole of Saturn's moon Enceladus on July 14, 2005, Cassini acquired high resolution views of this puzzling ice world. From afar, Enceladus exhibits a bizarre mixture of softened craters and complex, fractured terrains.

This graphic represents a possible model for mechanisms that could generate the water vapor and tiny ice particles detected by Cassini over the southern polar terrain on Enceladus. This model shows sublimation of an ammonia-water "slush" or "slurry" on the surface.

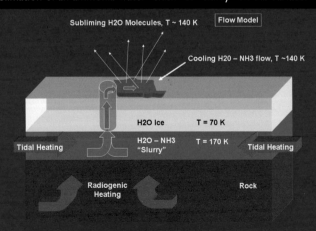

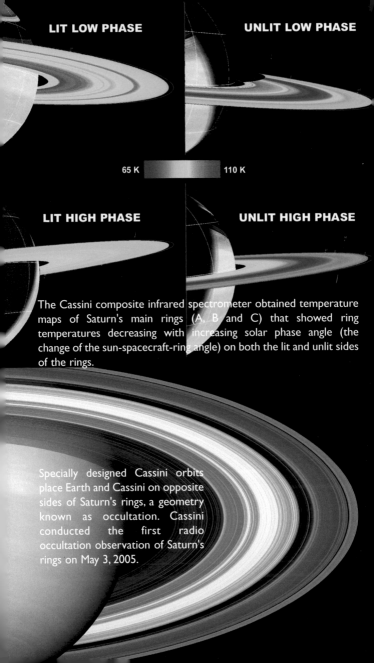

LIT LOW PHASE

UNLIT LOW PHASE

65 K 110 K

LIT HIGH PHASE

UNLIT HIGH PHASE

The Cassini composite infrared spectrometer obtained temperature maps of Saturn's main rings (A, B and C) that showed ring temperatures decreasing with increasing solar phase angle (the change of the sun-spacecraft-ring angle) on both the lit and unlit sides of the rings.

Specially designed Cassini orbits place Earth and Cassini on opposite sides of Saturn's rings, a geometry known as occultation. Cassini conducted the first radio occultation observation of Saturn's rings on May 3, 2005.

The dark Cassini Division, within Saturn's rings, contains a great deal of structure, as seen in this color image. The sharp inner boundary of the division (left of center) is the outer edge of the massive B ring and is maintained by the gravitational influence of the moon Mimas.

A large, bright and complex convective storm that appeared in Saturn's southern hemisphere in mid-September 2004 was the key in solving a long-standing mystery about the ringed planet. Saturn's atmosphere and its rings are shown here in a false color composite made from Cassini images taken in near infrared light through filters that sense different amounts of methane gas.

Tethys floats before the massive, golden-hued globe of Saturn in this natural color view. The thin, dark line of the rings curves around the horizon at top.

With its thick, distended atmosphere, Titan's orange globe shines softly, encircled by a thin halo of purple light-scattering haze. Small particles that populate high hazes in Titan's atmosphere scatter short wavelengths more efficiently than longer visible or infrared wavelengths, so the best possible observations of the detached layer are made in ultraviolet light.

Using visible light, astronomers detected clouds in the northern hemisphere of Uranus. These images, taken in the summer of 1997 by the Hubble Space Telescope's Wide Field and Planetary Camera 2, show banded structure and multiple clouds.

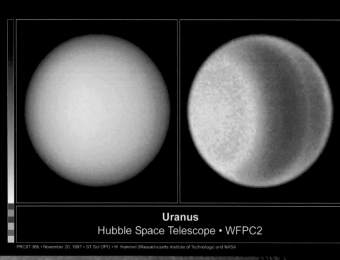

Uranus
Hubble Space Telescope • WFPC2

PRC97-36b • November 20, 1997 • ST ScI OPO • H. Hammel (Massachusetts Institute of Technology) and NASA

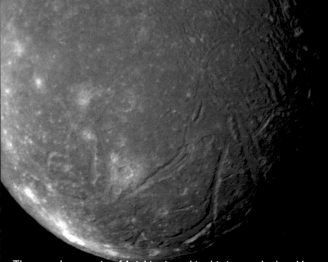

The complex terrain of Ariel is viewed in this image, the best Voyager 2 color picture of the Uranian moon. The individual photos used to construct this composite were taken January 24, 1986, from a distance of 170,000 kilometers (105,000 miles).

This computer generated montage shows Neptune as it would appear from a spacecraft approaching Triton, Neptune's largest moon at 2706 km (1683 mi) in diameter. Triton's surface is mostly covered by nitrogen frost mixed with traces of condensed methane, carbon dioxide, and carbon monoxide.

This image of Neptune was taken by Voyager 2's wide-angle camera when the spacecraft was 590,000 km (370,000 miles) from the planet. Processing allowed both the clouds' structure in the dark regions near the pole and the bright clouds east of the Great Dark Spot to be reproduced in this color photograph.

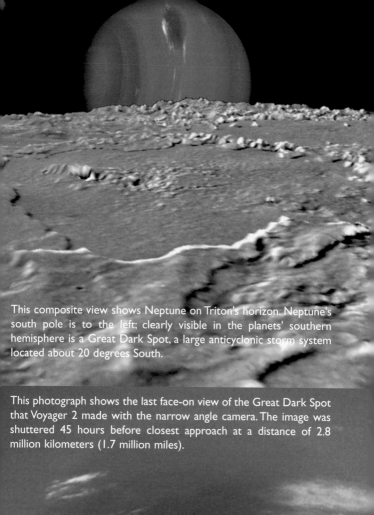

This composite view shows Neptune on Triton's horizon. Neptune's south pole is to the left; clearly visible in the planets' southern hemisphere is a Great Dark Spot, a large anticyclonic storm system located about 20 degrees South.

This photograph shows the last face-on view of the Great Dark Spot that Voyager 2 made with the narrow angle camera. The image was shuttered 45 hours before closest approach at a distance of 2.8 million kilometers (1.7 million miles).

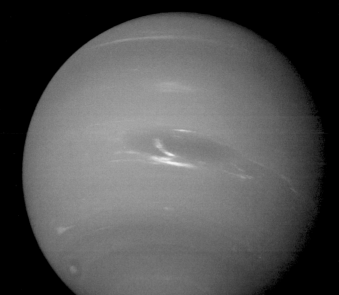

This picture of Neptune was produced from the last whole planet images taken through the green and orange filters on the Voyager 2 narrow angle camera. The images were taken at a range of 4.4 million miles from the planet, 4 days and 20 hours before closest approach.

This is a Hubble Telescope photo of Pluto and its moon Charon. Pluto is smaller than Earth's Moon and orbits at a tremendous distance from the Sun, so it is only a bright speck in the strongest telescopes.

Very recent Hubble images showed what may be two additional small moons of Pluto, bringing its total to three. As of January 2006, confirmation of these new moons is still pending.

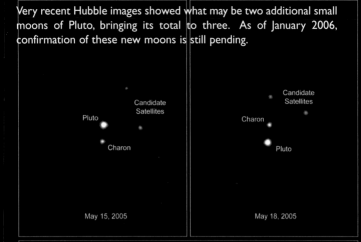

Reference Design Artist's Concept (April 2001) for SIM PlanetQuest, an optical interferometer operating in an Earth-trailing solar orbit.

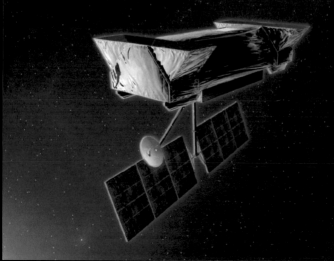

Artist's depiction of WMAP Leaving Earth/Moon Orbit, headed for the L2 Lagrange point, one million miles (1.5 million km) from Earth.

The WMAP Spacecraft was launched by Delta II Rocket on June 30, 2001.

In visible light, the bulk of our Milky Way galaxy's stars are eclipsed behind thick clouds of galactic dust and gas, but to the infrared eyes of NASA's Spitzer Space Telescope, distant stars and dust clouds shine with unparalleled clarity and color. In this panoramic image (center row), a plethora of stellar activity in the Milky Way's galactic plane is exposed.

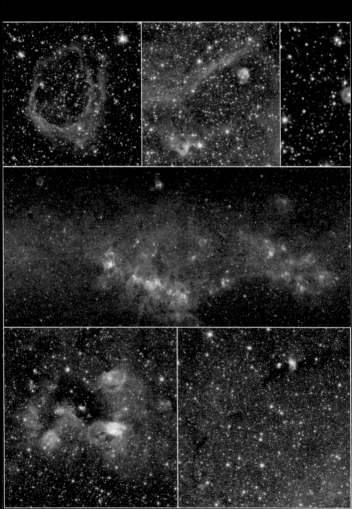

The red clouds indicate the presence of large organic molecules (mixed with the dust), which have been illuminated by nearby star formation. The patches of black are dense obscuring dust clouds impenetrable by even Spitzer's super-sensitive infrared eyes. Bright arcs of white throughout the image are massive stellar incubators.

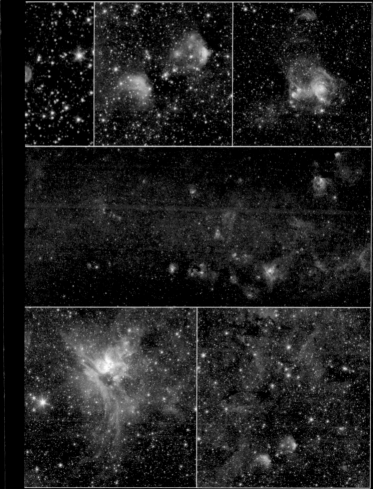

This infrared mosaic from NASA's Spitzer Space Telescope offers a stunning view of the stellar hustle and bustle that takes place at our Milky Way galaxy's center. The picture shows throngs of mostly old stars, on the order of hundreds of thousands, amid fantastically detailed clouds of glowing dust lit up by younger, massive stars.

The nearby galaxy NGC 2976, located approximately 10 million light-years away in the constellation Ursa Major near the Big Dipper, was captured by the Spitzer Infrared telescope.

ESA's Herschel Space Observatory will include the Spectral and Photometric Imaging Receiver (SPIRE), which will use "spider web bolometers," which are 40 times more sensitive than previous composite bolometers. SPIRE is being design and built by a consortium of institutes and university departments from across Europe, Canada and the USA.

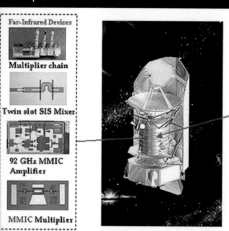

This is the central region of the Milky Way Galaxy as viewed in infrared light. The image is a composite of mid-infrared imagery from the MSX satellite and near-infrared imagery from the 2MASS survey. WISE images will be similar in quality.

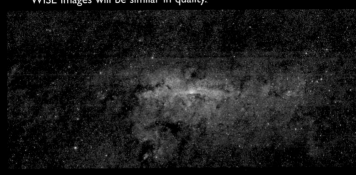

This is a composite image of visible and infrared light observations of Messier 104, commonly known as the Sombrero Galaxy. The visible light images were taken by the Hubble Space Telescope; the infrared observations were taken by the Spizter Space Telescope. The Sombrero Galaxy is a disk galaxy like the Milky Way and is located some 28 million light-years away. We view the disk of the galaxy nearly edge-on.

Launch of the Swift spacecraft from Cape Canaveral Air Force Station, November 20, 2004.

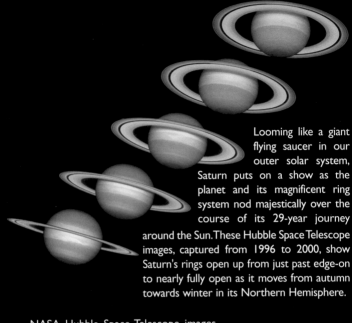

Looming like a giant flying saucer in our outer solar system, Saturn puts on a show as the planet and its magnificent ring system nod majestically over the course of its 29-year journey around the Sun. These Hubble Space Telescope images, captured from 1996 to 2000, show Saturn's rings open up from just past edge-on to nearly fully open as it moves from autumn towards winter in its Northern Hemisphere.

NASA Hubble Space Telescope images of the distant planet Neptune show a dynamic atmosphere and capture the fleeting orbits of its satellites. Images were taken in 14 different colored filters probing various altitudes in Neptune's deep atmosphere so that scientists can study the haze and clouds in detail.

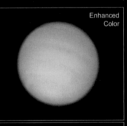

Enhanced Color

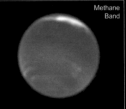

Methane Band

Neptune and Satellites
Hubble Space Telescope • ACS/HRC

This new Hubble image shows the most detailed view so far of the entire Crab Nebula ever made. The Crab is arguably the single most interesting object in all of astronomy. The image is the largest image ever taken with Hubble's WFPC2 workhorse camera.

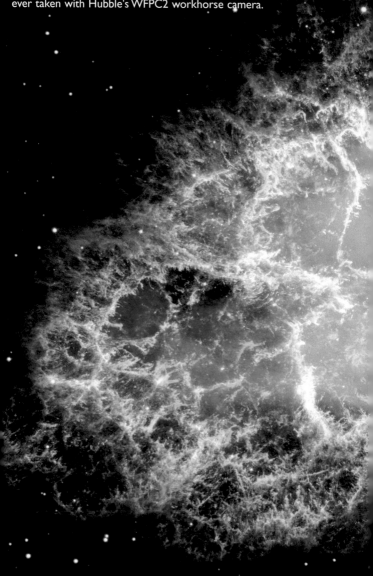

A Delta II launch vehicle carrying the X-ray Timing Explorer (XTE) lights up the early morning sky, December 30, 1995. The XTE spacecraft will study X-rays, including their origin and emission mechanisms, and the physical conditions and evolution of X-ray sources within the Milky Way galaxy and beyond.

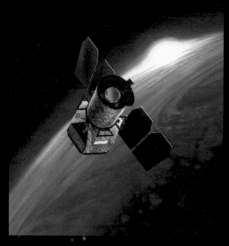

An artist's impression of GALEX in orbit around the Earth. The Solar panels are deployed and the telescope cover is open; nominal operations are proceeding.

A GALEX ultraviolet image of the barred spiral galaxy NGC 1365, which is a member of the Fornax Cluster of Galaxies.

Color Hubble Image of Multiple Comet Impacts on Jupiter.

This dramatic image offers a peek inside a cavern of roiling dust and gas where thousands of stars are forming. The image, taken by the Advanced Camera for Surveys (ACS) aboard NASA's Hubble Space Telescope, represents the sharpest view ever taken of this region, called the Orion Nebula. More than 3,000 stars of various sizes appear in this image.

Artist's rendering of the Stardust capsule's return to Earth. The Stardust spacecraft will bring back samples of interstellar dust, including recently discovered dust streaming into our Solar System from the direction of Sagittarius. The return date is January 15 2006.

The aerogel dust collector is the instrument aboard the Stardust spacecraft that captured interstellar dust. Each cell of the tennis racket-like detector holds aerogel in which the particles have become embedded.

NEW FROM THE #1 SPACE BOOK COMPANY –
APOGEE BOOKS

SPACE "POCKET REFERENCE GUIDES"

Each book has 96 pages with up to 48 pages of color. Packed with relevant data for each subject, such as crew photographs, flight statistics, time in space, distance traveled, mission objectives and a host of the best pictures taken during each program/mission.
www.apogeebooks.com